AF800865

# PERENNIAL MATH

# CONTEST PROBLEMS
## 2017-2018

Dr. Sylvia Dean was a gifted specialist and administrator for 26 years in Huntsville, Alabama. She created Perennial Math in 2011 upon completion of her doctoral degree. She has hosted many robotics, technology, and mathematics competitions throughout the years and has tirelessly worked on providing educational opportunities to students around the world. It has always been her ambition to leave students with the feeling that "math is cool."

These problems are challenging and students should know up front that scoring a 3 out of 6 on any of these tests is extraordinary and above average. Great problem-solvers learn persistence and appreciation for challenge. Good luck as you encourage your students to take the quest to become great mathletes.

# Rookie
## level

# Grades 3-4

Name________________________________

PERENNIAL MATH

# January 2017 Rookie Level

TEST #1

1) I am a 2-digit number. The sum of my digits is 12. The tens digit is three times as much as the ones digit. What number am I?

2) The auditorium has 450 seats. The 10 rows in front have 9 seats each. The rest of the rows have 15 seats each. How many rows have 15 seats?

3) Lisa and Katie counted the money that each had made from babysitting. Lisa said, "If I give you $10 we will have the same amount. If you give me $10, I will have 3 times as much as you." How much money does Lisa have?

4) In this magic square, the numbers 1 through 9 can be placed in the squares so that each row and column and diagonal add up to 15. What number goes in the top middle box?

| | | |
|---|---|---|
| | | 6 |
| 9 | | |
| | | 8 |

5) Matthew has $120. He plans to spend 10% of it on baseball cards and another 10% on socks. How much will money will he have left?

6) Emma has a box containing between 10 and 50 beads. If she counts them 2 at a time, she has one left over. If she counts them out 5 at a time she has 4 left over. If she counts them out 8 at a time she has 7 left over. How many beads does she have?

*Copyright © 2017 by Perennial Math, Inc. All rights reserved.*

PERENNIAL MATH

TEST #1

# January 2017 Rookie Level Answers

1) I am a 2-digit number. The sum of my digits is 12. The tens digit is three times as much as the ones digit. What number am I?

**93**

2) The auditorium has 450 seats. The 10 rows in front have 9 seats each. The rest of the rows have 15 seats each. How many rows have 15 seats?

**24**

3) Lisa and Katie counted the money that each had made from babysitting. Lisa said, "If I give you $10 we will have the same amount. If you give me $10, I will have 3 times as much as you." How much money does Lisa have?

**50**

4) In this magic square, the numbers 1 through 9 can be placed in the squares so that each row and column and diagonal add up to 15. What number goes in the top middle box?

| | | |
|---|---|---|
| | | 6 |
| 9 | | |
| | | 8 |

**7**

5) Matthew has $120. He plans to spend 10% of it on baseball cards and another 10% on socks. How much will money will he have left?

**96**

6) Emma has a box containing between 10 and 50 beads. If she counts them 2 at a time, she has one left over. If she counts them out 5 at a time she has 4 left over. If she counts them out 8 at a time she has 7 left over. How many beads does she have?

**39**

*Copyright © 2017 by Perennial Math, Inc. All rights reserved.*

**PERENNIAL MATH** TEST # 1

# January 2017 Rookie Level Solutions

1) To find a two-digit number whose sum of its digits is 12 and the tens digit is three times as much as the ones digit, you will need to list out numbers whose digits add to 12: 39, 48, 57, 66, 75, 84, 93. The answer is 93 because the tens digit is three times the ones digit.

**93**

2) The auditorium has 450 seats. The 10 rows in front have 9 seats each. The rest of the rows have 15 seats each. To solve for how many rows have 15 seats find the product of the first 10 rows and subtract it from the 450 seats. Then divide that amount by 15. 10 x 9 = 90 seats in the front. 450 – 90 = 360 seats. Now divide 360 by 5 and there are 24 rows that have 15 seats in each row.

**24**

3) Lisa said, "If I give you $10 we will have the same amount. If you give me $10, I will have 3 times as much as you." To solve for how much money Lisa has make a chart. Lisa has $50 because she and Katie will have the same amount if she gives Katie $10 and she will have three times Katie's amount if Katie gives her $10.

| Lisa | Katie | Lisa - $10 | Katie + $10 | Lisa + $10 | Katie - $10 |
|---|---|---|---|---|---|
| $20 | $10 | $10 | $20 | $30 | 0 |
| $30 | $10 | $20 | $20 | $40 | 0 |
| $40 | $20 | $30 | $30 | $50 | $10 |
| $50 | $30 | $40 | $40 | $60 | $20 |

**50**

*Copyright © 2017 by Perennial Math, Inc. All rights reserved.*

# PERENNIAL MATH

TEST # 1

## January 2017 Rookie Level Solutions

4) Since the numbers 1 through 9 can be placed in the squares so that each row and column and diagonal add up to 15. We can start solving for the unknowns to discover the number in the top middle box. First solve for the box between 6 and 8- it must be 1 so that the column equals 15. Next you can solve for the center box which must be 5. Now you can move to the top left corner and solve for the diagonal- this must be 2 since the diagonal sum is 8 + 5 + 2 = 15. Finally you can solve for the top middle box- It is 2 + ? + 6. It is 7.

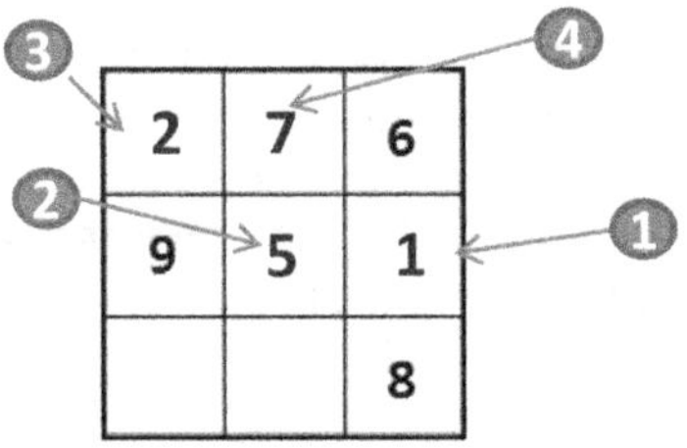


**7**

5) Matthew has $120. He plans to spend 10% of it on baseball cards and another 10% on socks. To find out how much will money will he have left first solve for what 20% of $120 is. Ten percent of $120 is $12 so twenty percent is $24. Now subtract this from the original amount $120 - $24 = $96.

**96**

6) Make a chart to find the first number that meets the requirements. The number is between 10 and 50 beads. If she counts them 2 at a time, she has one left over. This means her beads are divisible by 3. If she counts them out 5 at a time she has 4 left over. Start at 14 and count by 5. If she counts them out 8 at a time she has 7 left over. Start counting at 15 and count by 8. The number that is between 10 and 50 and that is meets all three counting requirements is 39. She has 39 beads.

| | | | | | | | | | |
|---|---|---|---|---|---|---|---|---|---|
| Counts by 2 with 1 left over | 15 | 18 | 21 | 24 | 27 | 30 | 33 | 36 | 39 |
| Counts by 5 with 4 left over | 14 | 19 | 24 | 29 | 34 | 39 | | | |
| Counts by 8 with 7 left over | 15 | 23 | 31 | 39 | | | | | |

**39**

*Copyright © 2017 by Perennial Math, Inc. All rights reserved.*

Name________________________________

PERENNIAL MATH

TEST #2

# February 2017 Rookie Level

1) (2016 + 2017) – (1998 + 1999) =

2) Candy A costs $6.00 a box. Candy B is $15.00 a box. How many boxes of Candy A are equal 10 boxes of Candy B?

3) In February Jason put some money in his bank account. Starting in March he put in $13 at the beginning of each month. On his birthday he put in an extra $85. By December 31st he had $325 in his account. How much money did Jason put in the bank to start with in February?

4) Following the lines, how many possible paths (without back tracking) are there from Point A to Point B?

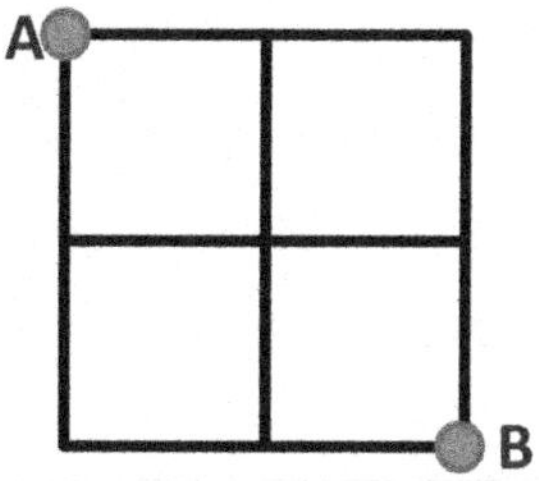


5) How many three-digit numbers can be made using 1, 2, 3, and 4 if you cannot repeat any of these numbers in the three-digit number?

6) At the zoo, there are three times as many monkeys as birds. There are also three zebras for every five monkeys. What is the smallest number of birds possible that could be at the zoo?

*Copyright © 2017 by Perennial Math, Inc. All rights reserved.*

# PERENNIAL MATH

## TEST #2

## February 2017 Rookie Level Answers

| | |
|---|---|
| 1) (2016 + 2017) – (1998 + 1999) = | 36 |
| 2) Candy A costs $6.00 a box. Candy B is $15.00 a box. How many boxes of Candy A are equal 10 boxes of Candy B? | 25 |
| 3) In February Jason put some money in his bank account. Starting in March he put in $13 at the beginning of each month. On his birthday he put in an extra $85. By December 31st he had $325 in his account. How much money did Jason put in the bank to start with in February? | 110 |
| 4) Following the lines, how many possible paths (without back tracking) are there from Point A to Point B? 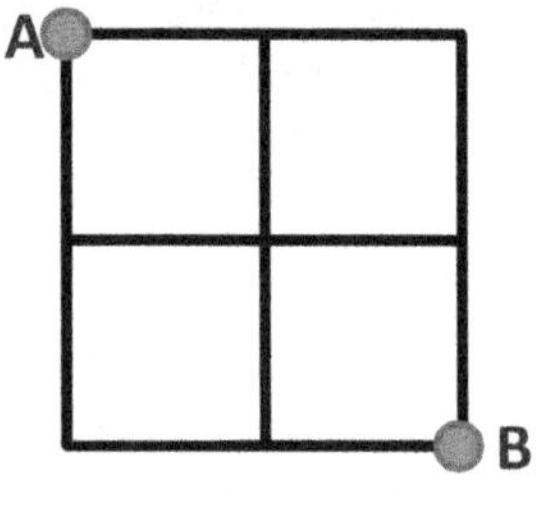  | 6 |
| 5) How many three-digit numbers can be made using 1, 2, 3, and 4 if you cannot repeat any of these numbers in the three-digit number? | 24 |
| 6) At the zoo, there are three times as many monkeys as birds. There are also three zebras for every five monkeys. What is the smallest number of birds possible that could be at the zoo? | 5 |

*Copyright © 2017 by Perennial Math, Inc. All rights reserved.*

PERENNIAL MATH TEST #2

# February 2017 Rookie Level Solutions

1) (2016 + 2017) – (1998 + 1999) = 4033 – 3997 = 36

**36**

2) To find out how many boxes of Candy A are equal 10 boxes of Candy B first find out how much 10 boxes of Candy B would cost and divide that by what the cost of Candy A $6.00. Candy B is $15.00 a box so 10 boxes is $15 x 10 = $150.  Now $150 ÷ $6 = 25 boxes.

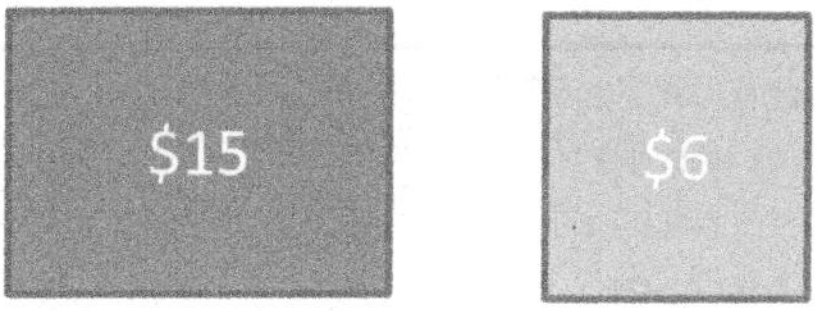


$15 x 10 = $150          $150 ÷ $6 = 25 boxes

**25**

3) In February Jason put some money in his bank account. After that he put in $13 at the beginning of each month. So count up the months from March through December and there are 10 months. Ten times $13 = $130.  On his birthday he put in an extra $85. So $130 + $85 = $215. By December 31st he had $325 in his account. Subtract the money he put in all year $215 and you will end up with the amount of money he first put in the bank in February. $325 - $215 = $110

**110**

*Copyright © 2017 by Perennial Math, Inc. All rights reserved.*

# PERENNIAL MATH

TEST #2

## February 2017 Rookie Level Solutions

4) Draw out the pattern and keep track of the directions you took so that you don't repeat yourself. Try to stick to a consistent method, such as going across the top and getting all of the paths that way, then moving to the middle and getting all the paths that way, and then finally to the bottom line. There are 6 separate paths.

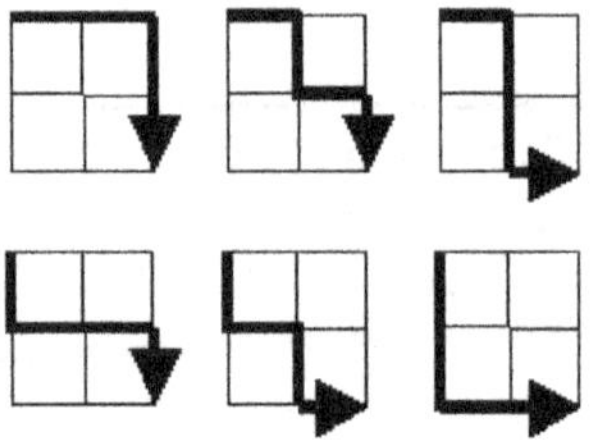

**6**

5) Make an organized list. Start with 1 and make the numbers in order that can be made and do the same for each column. There are 6 numbers that can be made that begin with each number for a total of 24 numbers.

| Numbers beginning with 1 | Numbers beginning with 2 | Numbers beginning with 3 | Numbers beginning with 4 |
|---|---|---|---|
| 123 | 213 | 312 | 412 |
| 124 | 214 | 314 | 413 |
| 132 | 231 | 321 | 421 |
| 134 | 234 | 324 | 423 |
| 142 | 241 | 341 | 431 |
| 143 | 243 | 342 | 432 |

**24**

6) Make a chart to keep up with the requirements. There are three times as many monkeys as birds. There are also three zebras for every five monkeys. The number of monkeys is a multiple of 5. Since there are three times as many monkeys as birds, the first multiple of 5 that can be three times as much is 15. Since there is a ratio of 5 to 3 for monkeys to zebras 15:9 is a 5:3 ratio. The smallest number of birds possible at the zoo is 5.

| Birds | Monkeys | Zebras |
|---|---|---|
| 1 | 3 | Not possible |
| 2 | 6 | Not possible |
| 5 | 15 | 9 |

**5**

*Copyright © 2017 by Perennial Math, Inc. All rights reserved.*

Name________________________________

PERENNIAL MATH

**March 2017 Rookie Level**

**TEST #3**

| | |
|---|---|
| **1) _____+ 146 + 379 + 284 = 1173** | |
| **2) The leprechaun grew one four-leaf clover for every ten three-leaf clovers he grew. He counted 44 clovers in his garden. How many total leaves did he count?** | |
| **3) There are three times as many boys as girls in a room . If 6 boys leave the room, there would be an equal number of boys and girls in the room. How many boys and girls altogether were in the room at first?** | |
| **4) The perimeter of a rectangle is 48 inches. The length of the rectangle is 18 inches. What is the area of the rectangle?** 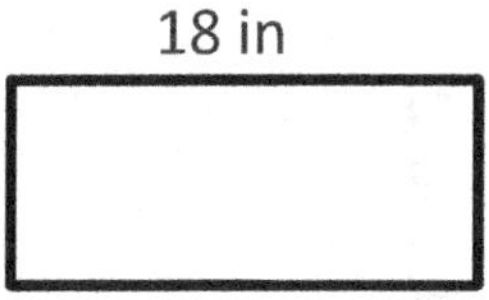  | |
| **5) A bike museum has a collection of bicycles and tricycles. Together there are 36 pedals and 48 wheels. How many tricycles are in the museum?** | |
| **6) From the digits 1 through 9 how many three-digit numbers can be created that contain a four and have a sum of 14 if digits can be repeated?** | |

*Copyright © 2017 by Perennial Math, Inc. All rights reserved.*

PERENNIAL MATH TEST #3

# March 2017 Rookie Level Answers

| | |
|---|---|
| 1) _____+ 146 + 379 + 284 = 1173 | **364** |
| 2) The leprechaun grew one four-leaf clover for every ten three-leaf clovers he grew. He counted 44 clovers in his garden. How many total leaves did he count? | **136** |
| 3) There are three times as many boys as girls in a room. If 6 boys leave the room, there would be an equal number of boys and girls in the room. How many boys and girls altogether were in the room at first? | **12** |
| 4) The perimeter of a rectangle is 48 inches. The length of the rectangle is 18 inches. What is the area of the rectangle? 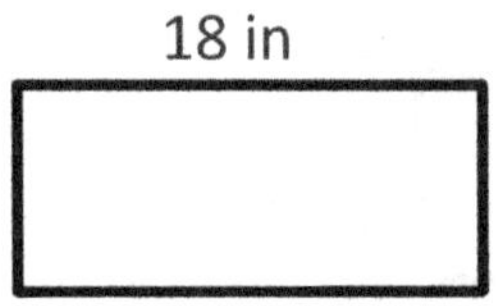  | **108** |
| 5) A bike museum has a collection of bicycles and tricycles. Together there are 36 pedals and 48 wheels. How many tricycles are in the museum? | **12** |
| 6) From the digits 1 through 9 how many three-digit numbers can be created that contain a four and have a sum of 14 if digits can be repeated? | **24** |

*Copyright © 2017 by Perennial Math, Inc. All rights reserved.*

PERENNIAL MATH TEST #3

# March 2017 Rookie Level Solutions

1) _____ + 146 + 379 + 284 = 1173.

Add the three numbers together and subtract from the answer to find out the missing number:

$$\begin{array}{r} 146 \\ 379 \\ \underline{+284} \\ 809 \end{array}$$

1173 - 809 = 364

**364**

2) Since the leprechaun grew one four-leaf clover for every ten three-leaf clovers he grew the ratio is 10:1. He counted 44 clovers in his garden. So that was a ratio of 40:4 for three-leaf to four-leaf clovers. 40 x 3 = 120 leaves plus the 4 x 4 = 16 leaves is a total of 136 leaves that he counted.

**136**

3) There are three times as many boys as girls in a room . If 6 boys leave the room, there would be an equal number of boys and girls in the room. How many boys and girls altogether were in the room at first?

The ratio of boys to girls was 3:1. When the 6 boys left there were an equal number of boys and girls. So 6 is equal to 2 portions. So a portion is equal to 3. There are 12 students altogether at first.

**12**

Copyright © 2017 by Perennial Math, Inc. All rights reserved.

PERENNIAL MATH

TEST #3

# March 2017 Rookie Level Solutions

4) The perimeter of a rectangle is 48 inches. The length of the rectangle is 18 inches. To solve for the width multiply the length by 2 and subtract this amount from the perimeter. 48 – 2(18) = 48 – 36 = 12 inches. Now divide 12 inches by 2 to solve for what the width sides must equal. The width is 6 inches. Area is equal to length times width so the area is 18in. x 6in. = 108 $in^2$.

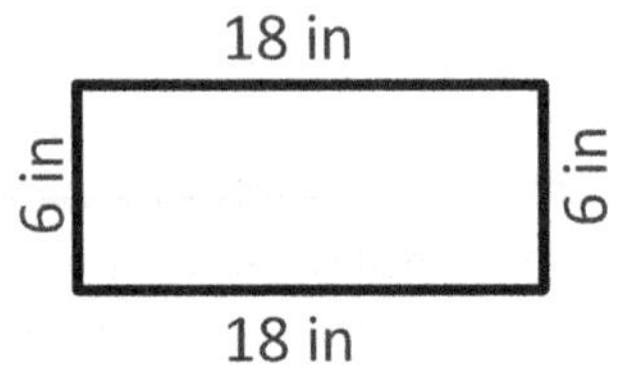


**108**

5) A bike museum has a collection of bicycles and tricycles and there are 36 pedals and 48 wheels. Since there are 36 pedals there must be 18 total vehicles because each has 2 pedals. Create a table to find out how many tricycles are in the museum. If you start with an equal number of each you will see that you need to increase the number of tricycles. There are 12 tricycles in the museum.

| | | | | |
|---|---|---|---|---|
| Bikes | 9 | 8 | 7 | 6 |
| Tricycles | 9 | 10 | 11 | 12 |
| Total Pedal | 36 | 36 | 36 | 36 |
| Total Wheels | 45 | 46 | 47 | 48 |

**12**

6) To find out the three-digit numbers that can be made using the digits 1-9 make an organized list starting with each number and adding 4 to it to see what three-digit combinations are possible that add to 14. There are 24 numbers that can be made.

| | | | | | | |
|---|---|---|---|---|---|---|
| All arangements with 1, 4, 9 | 149 | 194 | 419 | 491 | 914 | 941 |
| All arangements with 2, 4, 8 | 248 | 284 | 428 | 482 | 824 | 842 |
| All arangements with 3, 4, 7 | 347 | 374 | 437 | 473 | 734 | 743 |
| numbers with double fours | 446 | 464 | 644 | | | |
| numbers with double fives | 455 | 545 | 554 | | | |

**24**

*Copyright © 2017 by Perennial Math, Inc. All rights reserved.*

Name________________________________

PERENNIAL MATH

TEST #4

## April 2017 Rookie Level

| | |
|---|---|
| 1) 51 + 52 + 53 + 54 + 55 + 56 + 57 + 58 + 59 = | |
| 2) Chris gave away 48 of his marbles on the first day. On the second day he gave away 22 less than the first day. On the third day he gave away double what he gave away on the first day. And on the last day he gave away 58 more than he gave away on the third day. How many marbles did he give away in all? | |
| 3) The sum of Morris and Mike's age is 48 years. Morris is twice Mike's age. How old is Mike? | |
| 4) How many triangles are in the picture below?  | |
| 5) My computer is only capable of one process: When I enter a two-digit number, it reverses its digits, subtracts 7 from the result, and gives this final number as its answer. When I put in certain number, the computer gave me 36 as its final answer. What number did I enter? | |
| 6) A grandfather clock chimes every hour and the number of chimes is equal to the hour it chimes. For instance at 4:00 it chimes 4 times. How many times will it chime in one week? | |

Copyright © 2017 by Perennial Math, Inc. All rights reserved.

# PERENNIAL MATH

TEST #4

## April 2017 Rookie Level Answers

| | |
|---|---|
| 1) 51 + 52 + 53 + 54 + 55 + 56 + 57 + 58 + 59 = | **495** |
| 2) Chris gave away 48 of his marbles on the first day. On the second day he gave away 22 less than the first day. On the third day he gave away double what he gave away on the first day. And on the last day he gave away 58 more than he gave away on the third day. How many marbles did he give away in all? | **324** |
| 3) The sum of Morris and Mike's age is 48 years. Morris is twice Mike's age. How old is Mike? | **16** |
| 4) How many triangles are in the picture below?  | **8** |
| 5) My computer is only capable of one process: When I enter a two-digit number, it reverses its digits, subtracts 7 from the result, and gives this final number as its answer. When I put in certain number, the computer gave me 36 as its final answer. What number did I enter? | **34** |
| 6) A grandfather clock chimes every hour and the number of chimes is equal to the hour it chimes. For instance at 4:00 it chimes 4 times. How many times will it chime in one week? | **1,092** |

*Copyright © 2017 by Perennial Math, Inc. All rights reserved.*

PERENNIAL MATH

TEST #4

# April 2017 Rookie Level Solutions

1) 51 + 52 + 53 + 54 + 55 + 56 + 57 + 58 + 59 = 495
Or use smart addition. Group the ones: 1 + 9, 2 + 8, 3 + 7, 4 + 6, + 5 = 45
Now add up the tens: 50 x 9 = 450
450 + 45 = 495

**495**

2) Make a chart to keep up with what Chris gave away each day and then add up the amounts. Chris gave away 48 of his marbles on the first day. On the second day he gave away 22 less than the first day (48 – 22 = 26). On the third day he gave away double what he gave away on the first day (48 x 2= 96). And on the last day he gave away 58 more than he gave away on the third day (96 + 58 = 154) .

| | Procedure | Total Given Away |
|---|---|---|
| Day 1 | | 48 |
| Day 2 | 48 - 22 | 26 |
| Day 3 | 48 x 2 | 96 |
| Day 4 | 96 + 58 | 154 |
| Total | | 324 |

**324**

3) Make a chart to find out Mike's age. The sum of Morris and Mike's age is 48 years. Morris is twice Mike's age. If you create a chart you will see that each time you raise Mike's age 1 year, the sum of their ages goes up 3 year. Mike is 16 years old.

| Mike | Morris (two times Mike) | Sum (Must = 48) |
|---|---|---|
| 10 | 20 | 30 |
| 11 | 22 | 33 |
| 12 | 24 | 36 |
| 13 | 26 | 39 |
| 14 | 28 | 42 |
| 15 | 30 | 45 |
| 16 | 32 | 48 |

*Increase of 3 years*

**16**

*Copyright © 2017 by Perennial Math, Inc. All rights reserved.*

PERENNIAL MATH TEST #4

## April 2017 Rookie Level Solutions

4) Count the individual triangles and triangles that are combination triangles. There are 8 different triangles.

**8**

5) Since the computer reverses the digits and then subtracts 7 from the to give its answer you will need to work backwards to find the original number entered with 36 as a final answer. Add 7 to 36 = 43. Now reverse the numbers to 34.

**34**

6) To solve how many times a clock will chime in one week, first find out how many times it will chime in 1 day and then multiply that amount by 7.

The first 12 hours it chimes: 1 + 2 + 3 + 4 + 5 + 6 +7 + 8 + 9 + 10 + 11 + 12 = 78 times.

So it does this twice per day since there are 24 hours in a day. 78 x 2 = 156 times chimes per day.

Multiply this by 7 to solve for chimes in one week. 156 x 7 = 1,092 chimes.

**1,092**

*Copyright © 2017 by Perennial Math, Inc. All rights reserved.*

Name________________________________

**PERENNIAL MATH**

**TEST #1**

## November 2017 Rookie Level

1) 292 – 57 = ________ + 138

2) Bob called his aunt to tell her he won the spelling bee. His aunt called 4 other people to tell them the news. Each of these four people called 4 other people who called 4 other people each. So how many people ended up hearing the news about Bob (do not include Bob)?

3) Mike and Larry are moving and they are taking their large video collection with them. They packed 17 videos in each box. They put 5 boxes on each of the 7 tables. However, their neighbor came by and they gave him 15 of the videos. How many total videos did they have left?

4) The groups of pattern blocks are placed in rows with the lengths of each row given. Given the information below, how many centimeters are in the width of one diamond?

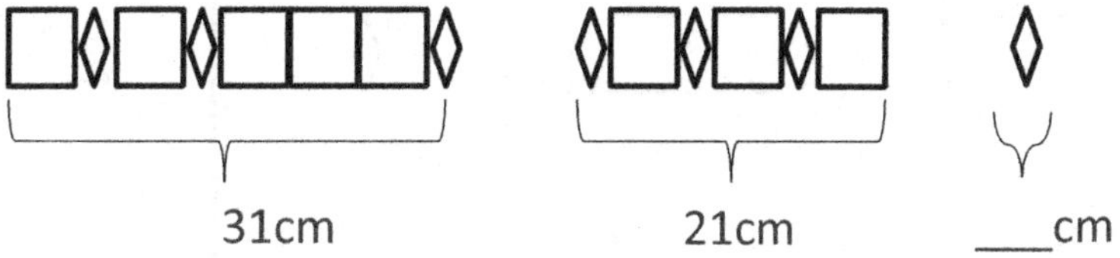


5) Jack had 72 marbles. He gave half his marbles to his brother. He then gave an equal number to each of his four friends. He now has 12 marbles. How many marbles did he give to each of his friends?

6) How many ways can 16 be written as the sum of 3 different numbers?

*Copyright © 2017 by Perennial Math, Inc. All rights reserved.*

# PERENNIAL MATH

**TEST #1**

## November 2017 Rookie Level Answers

| | |
|---|---|
| 1) 292 – 57 = _______ + 138 | **97** |
| 2) Bob called his aunt to tell her he won the spelling bee. His aunt called 4 other people to tell them the news. Each of these four people called 4 other people who called 4 other people each. So how many people ended up hearing the news about Bob (do not include Bob)? | **85** |
| 3) Mike and Larry are moving and they are taking their large video collection with them. They packed 17 videos in each box. They put 5 boxes on each of the 7 tables. However, their neighbor came by and they gave him 15 of the videos. How many total videos did they have left? | **580** |
| 4) The groups of pattern blocks are placed in rows with the lengths of each row given. Given the information below, how many centimeters are in the width of one diamond? 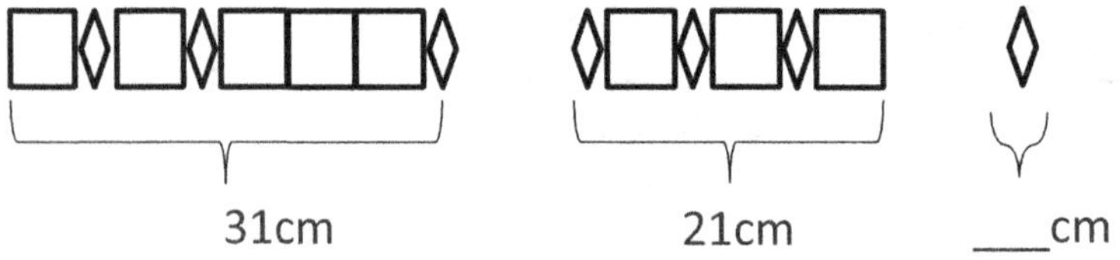  | **2** |
| 5) Jack had 72 marbles. He gave half his marbles to his brother. He then gave an equal number to each of his four friends. He now has 12 marbles. How many marbles did he give to each of his friends? | **6** |
| 6) How many ways can 16 be written as the sum of 3 different numbers? | **14** |

*Copyright © 2017 by Perennial Math, Inc. All rights reserved.*

PERENNIAL MATH

TEST #1

# November 2017 Rookie Level Solutions

1) 292 – 57 = _______ + 138

Solve for the left side of the equation first. 292-57 = 235. Now subtract 138 from the answer of that equation so that you know what is needed to make the right side equivalent. 235 – 138 = 97. The missing number is 97.

**97**

2) Make a diagram to keep up with how many people were called. Bob called his aunt. His aunt called 4 other people. Each of these four people called 4 other people who called 4 other people each. So the number of people called is 1 + 4 + 16 + 64 = 85 people.

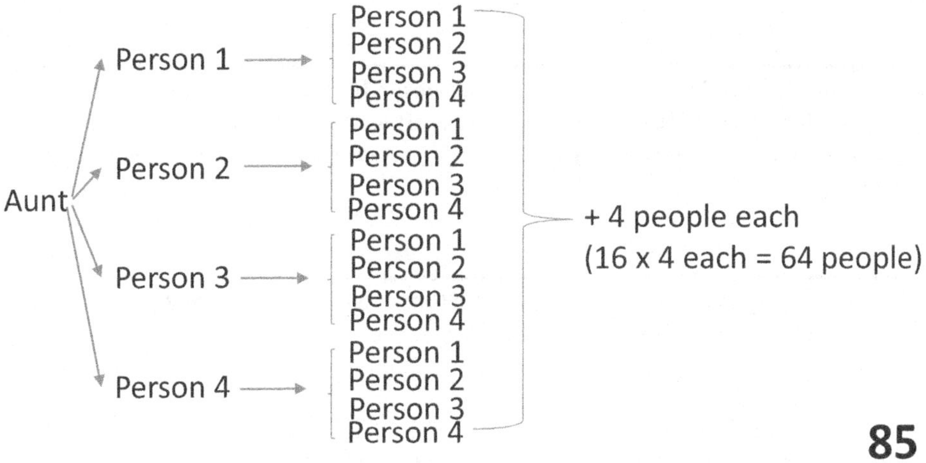


**85**

3) Since they packed 17 videos in each box and they put 5 boxes on each of the 7 tables, you can use multiplication to solve for how many videos they started with. 17 x 5 x 7 = 595. Now subtract the 15 videos they gave the neighbor. 595 – 15 = 580 videos that they have left.

**580**

*Copyright © 2017 by Perennial Math, Inc. All rights reserved.*

# PERENNIAL MATH

TEST # 1

## November 2017 Rookie Level Solutions

4) Given the information below, we can find how many centimeters are in the width of one diamond. 3 diamonds plus 5 squares = 31cm. 3 diamonds plus 3 squares = 21 cm. Subtract the three diamonds and three squares out from the 31cm and you are left with 2 squares = 10cm. So one square = 5 cm. Now Solve for the diamond by plugging in the 5cm for each square. 3 diamonds plus 3 squares (3 x 5cm) = 21 cm. 21cm – 15cm = 6cm. So if three diamonds = 6cm, then one diamond = 6cm / 3 = 2cm.

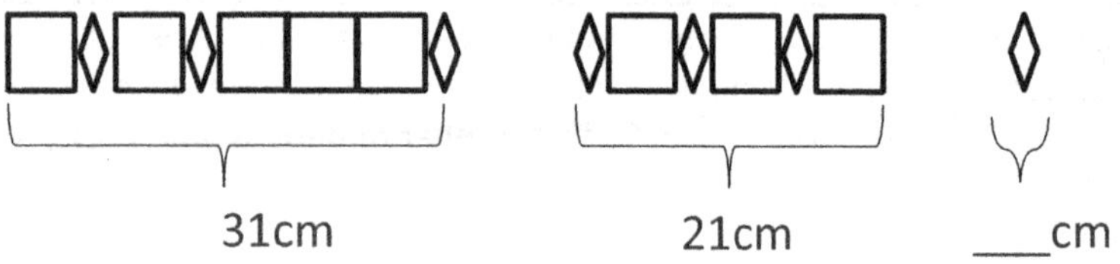


**2**

5) Jack had 72 marbles. He gave half his marbles to his brother. So then Jack had 72 / 2 = 36 marbles. 36 – 4x = 12. 36 marbles – 12 marbles = 24 marbles. The amount that he gave all four of his friends (4x) is 24. 24/4 = 6 marbles each. He gave each of his friends 6 marbles.

**6**

6) How many ways can 16 be written as the sum of 3 different numbers? Start with lowest to highest and make an organized list.

| | | | |
|---|---|---|---|
| 1 + 2 + 13 | 2 + 3 + 11 | 3 + 4 + 9 | 4 + 5 + 7 |
| 1 + 3 + 12 | 2 + 4 + 10 | 3 + 5 + 8 | |
| 1 + 4 + 11 | 2 + 5 + 9 | 3 + 6 + 7 | |
| 1 + 5 + 10 | 2 + 6 + 8 | | |
| 1 + 6 + 9 | | | |
| 1 + 7 + 8 | | | |

**14**

*Copyright © 2017 by Perennial Math, Inc. All rights reserved.*

Name________________________________

# PERENNIAL MATH

**TEST #2**

## December 2017 Rookie Level

1) In gym class there are 74 boys and 59 girls. Yesterday 4 boys and 2 girls were absent. So there were _____ more boys than girls in the class yesterday.

2) Harmony Record Shop sold 16 CDs for $12 each. If they keep ¼ of the money as profit, how much profit do they keep?

3) Apples cost 37¢, bananas cost 27¢, and kiwis cost 59¢. How much more does it cost for two kiwis and 1 banana than 2 bananas and 1 apple? Write your answer as a decimal.

4) In this pyramid the number in each block is the sum of the two numbers directly below it. What number goes in the block with the ?

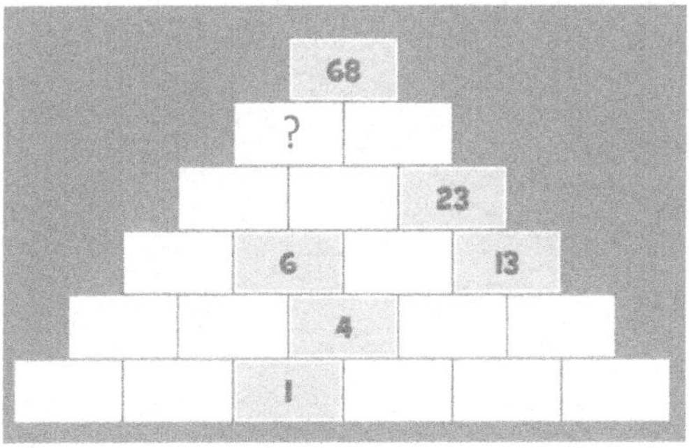


5) If you start at 300 and start counting backwards by 7 and then by 17 on a rotating basis, what is the last number you will count before you reach zero? For example 300, 293, 276...

6) Billy has pennies, nickels, dimes, and quarters. How many ways can he make 47 cents?

*Copyright © 2017 by Perennial Math, Inc. All rights reserved.*

PERENNIAL MATH

TEST #2

# December 2017 Rookie Level Answers

| | |
|---|---|
| 1) In gym class there are 74 boys and 59 girls. Yesterday 4 boys and 2 girls were absent. So there were _____ more boys than girls in the class yesterday. | 13 |
| 2) Harmony Record Shop sold 16 CDs for $12 each. If they keep ¼ of the money as profit, how much profit do they keep? | 48 |
| 3) Apples cost 37¢, bananas cost 27¢, and kiwis cost 59¢. How much more does it cost for two kiwis and 1 banana than 2 bananas and 1 apple? Write your answer as a decimal. | .54 |
| 4) In this pyramid the number in each block is the sum of the two numbers directly below it. What number goes in the block with the ? 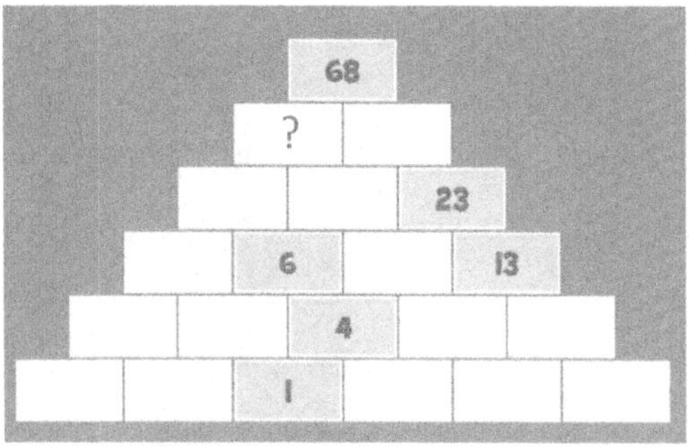  | 29 |
| 5) If you start at 300 and start counting backwards by 7 and then by 17 on a rotating basis, what is the last number you will count before you reach zero? For example 300, 293, 276... | 5 |
| 6) Billy has pennies, nickels, dimes, and quarters. How many ways can he make 47 cents? | 39 |

*Copyright © 2017 by Perennial Math, Inc. All rights reserved.*

# PERENNIAL MATH

TEST #2

## December 2017 Rookie Level Solutions

1) In gym class there are 74 boys and 59 girls. Yesterday 4 boys and 2 girls were absent. So there were _____ more boys than girls in the class yesterday. Subtract the boys who were absent: 74 – 4 = 70 boys present. Subtract the girls who were absent: 59 – 2 = 57 girls present. Now subtract the girls from the boys: 70-57 = 13. There were 13 more boys present than girls.

**13**

2) Harmony Record Shop sold 16 CDs for $12 each. They keep ¼ of the money as profit. To find how much profit they keep first solve for how much total money they make and then divide it by 4. 16 x $12 = $192. $192/4 = $48. They made a profit of $48.

**48**

3) Apples cost 37¢, bananas cost 27¢, and kiwis cost 59¢. How much more does it cost for two kiwis and 1 banana than 2 bananas and 1 apple? Write your answer as a decimal. Two kiwis and 1 banana costs 2(.59) + .27 = 1.18 + .27 = 1.45. Two bananas and 1 apple costs 2(.27) + .37 = .54 + .37 =.91. Now subtract the amounts to find out how much more the two kiwis and 1 banana cost: 1.45 - .91 = .54

**.54**

*Copyright © 2017 by Perennial Math, Inc. All rights reserved.*

# PERENNIAL MATH

TEST #2

## December 2017 Rookie Level Solutions

4) Solve the puzzle by looking for numbers to start at that will help you get to the missing number "?". If you look closely, the block between the original 6 and 13 is a great starting point. Fill in the missing numbers and work your way to the up. Once you solve to the 39 in the top portion you can figure out that 39 + ? = 68. The missing number is 29.

**29**

5) You can start and count backwards, or divide by 24 and see what the remainder would be. 300, 293, 276, 269, 252, 245, 228, 221, 204, 197, 180, 173, 156, 149, 132, 125, 108, 101, 84, 77, 60, 53, 36, 29, 12, 5.

**5**

6) Billy has pennies, nickels, dimes, and quarters. To find how many ways he can make 47 cents make an organized list: There are 39 ways to make 47 cents.

| Q | D | N | P |
|---|---|---|---|
| 1 | 2 | 0 | 2 |
| 1 | 1 | 2 | 2 |
| 1 | 1 | 1 | 7 |
| 1 | 1 | 0 | 12 |
| 1 | 0 | 4 | 2 |
| 1 | 0 | 3 | 7 |
| 1 | 0 | 2 | 12 |
| 1 | 0 | 1 | 17 |
| 1 | 0 | 0 | 22 |
| 0 | 4 | 1 | 2 |
| 0 | 4 | 0 | 7 |
| 0 | 3 | 3 | 2 |
| 0 | 3 | 2 | 7 |

| Q | D | N | P |
|---|---|---|---|
| 0 | 3 | 1 | 12 |
| 0 | 3 | 0 | 17 |
| 0 | 2 | 5 | 2 |
| 0 | 2 | 4 | 7 |
| 0 | 2 | 3 | 12 |
| 0 | 2 | 2 | 17 |
| 0 | 2 | 1 | 22 |
| 0 | 2 | 0 | 27 |
| 0 | 1 | 7 | 2 |
| 0 | 1 | 6 | 7 |
| 0 | 1 | 5 | 12 |
| 0 | 1 | 4 | 17 |
| 0 | 1 | 3 | 22 |

| Q | D | N | P |
|---|---|---|---|
| 0 | 1 | 2 | 27 |
| 0 | 1 | 1 | 32 |
| 0 | 1 | 0 | 37 |
| 0 | 0 | 9 | 2 |
| 0 | 0 | 8 | 7 |
| 0 | 0 | 7 | 12 |
| 0 | 0 | 6 | 17 |
| 0 | 0 | 5 | 22 |
| 0 | 0 | 4 | 27 |
| 0 | 0 | 3 | 32 |
| 0 | 0 | 2 | 37 |
| 0 | 0 | 1 | 42 |
| 0 | 0 | 0 | 47 |

**39**

*Copyright © 2017 by Perennial Math, Inc. All rights reserved.*

Name________________________________

PERENNIAL MATH

TEST #3

## January 2018 Rookie Level

1) Rosie gets paid $7 per hour for filing at her mom's office. If she works from 8:30 a.m. to 6:30 p.m., then how much money will she make?

2) Vera's pool is 12 meters long, so she swims 24 meters when she swims a lap. If she swims 6 laps a day for an entire week, but her brother swims 7 and a half laps a day for a week, how many more meters is her brother swimming than her for the week?

3) Tani brought 17 trays of cookies with 14 cookies on each tray. Jill brought 13 trays of 15 cookies each. How many more cookies did Tani bring?

4) The problem below is a cryptarithm. The letters A and B stand for different digits. What three-digit number would the letters CBB represent in this addition problem?

$$\begin{array}{r} AB \\ +\ AB \\ \hline CBB \end{array}$$

5) What three numbers added together equal 15 and multiplied together equal 105? Write your answer with the numbers going highest to lowest. (example 321)

6) If one circle is worth three stars, and one star is worth eight triangles, how many circles are seventy-two triangles worth?

Copyright © 2018 by Perennial Math, Inc. All rights reserved.

# PERENNIAL MATH

TEST #3

## January 2018 Rookie Level Answers

| | |
|---|---|
| 1) Rosie gets paid $7 per hour for filing at her mom's office. If she works from 8:30 a.m. to 6:30 p.m., then how much money will she make? | 70 |
| 2) Vera's pool is 12 meters long, so she swims 24 meters when she swims a lap. If she swims 6 laps a day for an entire week, but her brother swims 7 and a half laps a day for a week, how many more meters is her brother swimming than her for the week? | 252 |
| 3) Tani brought 17 trays of cookies with 14 cookies on each tray. Jill brought 13 trays of 15 cookies each. How many more cookies did Tani bring? | 43 |
| 4) The problem below is a cryptarithm. The letters A and B stand for different digits. What three-digit number would the letters CBB represent in this addition problem? $\begin{array}{r} AB \\ +\ AB \\ \hline CBB \end{array}$ | 100 |
| 5) What three numbers added together equal 15 and multiplied together equal 105? Write your answer with the numbers going highest to lowest. (example 321) | 753 |
| 6) If one circle is worth three stars, and one star is worth eight triangles, how many circles are seventy-two triangles worth? | 3 |

*Copyright © 2018 by Perennial Math, Inc. All rights reserved.*

# PERENNIAL MATH

TEST # 3

## January 2018 Rookie Level Solutions

1) Rosie gets paid $7 per hour for filing at her mom's office. If she works from 8:30 a.m. to 6:30 p.m., then how much money will she make? First solve for how much time she worked and then multiple that by her hourly rate. She worked 10 hours. So, 10 x $7 = $70.

**70**

2) Vera's pool is 12 meters long, so she swims 24 meters when she swims a lap. Vera swims 6 laps a day, but her brother swims 7 and a half laps a day for the week. To find out how many more meters her brother swims for the week, first solve for how many more laps he swims. He swims 1.5 more laps a day x 7 days = 10.5 more laps. Each lap is 24 meters. So 24 meters x 10.5 = 252 more meters.

**252**

3) Tani brought 17 trays of cookies with 14 cookies on each tray. Jill brought 13 trays of 15 cookies each. To find how many more cookies Tani brought find out the amount each girl brought and subtract. Tani brought 17 x 14 = 238 cookies. Jill brought 13 x 15 =195. 238 – 195 = 43.

**43**

*Copyright © 2018 by Perennial Math, Inc. All rights reserved.*

# PERENNIAL MATH

TEST # 3

## January 2018 Rookie Level Solutions

4) What three-digit number would the letters CBB represent in this addition problem? Since B + B = B in the units column, then B can only be 0,. Also notice that A + A carries, so A must be 5 or higher. In order for A + A to yield B or (0), A must equal 5. The number CBB stands for is 100.

$$\begin{array}{r} AB \\ +\ AB \\ \hline CBB \end{array}$$

**100**

5) Start with an organized list to find three numbers with sums of 15 and products of 105. Arrange the numbers highest to lowest. The answer is 753.

| Number | Number | Number | Product |
|---|---|---|---|
| 1 | 1 | 13 | 13 |
| 1 | 2 | 12 | 24 |
| 1 | 3 | 11 | 33 |
| 1 | 4 | 10 | 40 |
| 1 | 5 | 9 | 45 |
| 1 | 6 | 8 | 48 |
| 1 | 7 | 7 | 49 |
| 2 | 2 | 11 | 44 |
| 2 | 3 | 10 | 60 |
| 2 | 4 | 9 | 72 |
| 2 | 5 | 8 | 80 |
| 2 | 6 | 7 | 84 |
| 3 | 3 | 9 | 81 |
| 3 | 4 | 8 | 96 |
| 3 | 5 | 7 | 105 |
| 3 | 6 | 6 | 108 |
| 4 | 4 | 7 | 112 |
| 4 | 5 | 6 | 120 |
| 5 | 5 | 5 | 125 |

**753**

6) One circle is worth three stars. One star is worth eight triangles. You can draw this out to see the relationship of ratios and how many circles 72 triangles are worth. 72 triangles divided by 8 are worth 9 stars which would be equal to 3 circles.

**3**

*Copyright © 2018 by Perennial Math, Inc. All rights reserved.*

Name________________________________

PERENNIAL MATH

TEST #4

## February 2018 Rookie Level

1) Two numbers when added together equal 20 and when subtracted equal 6. What is the largest of the two numbers?

2) Gumballs cost 7 cents and chocolate drops cost 16 cents. If you had to purchase at least one of each, what is the least amount of money you would have to spend if you spent more than 75 cents?

3) Matthew gave 6 of his 30 marbles to Carl. Now they both have the same number of marbles. They both gave away one fourth of their marbles to Eric. How many marbles did Eric receive?

4) How many ways can a ladybug crawl from the bottom right corner to the top left corner of this figure if it can only travel up, left or both along the lines?

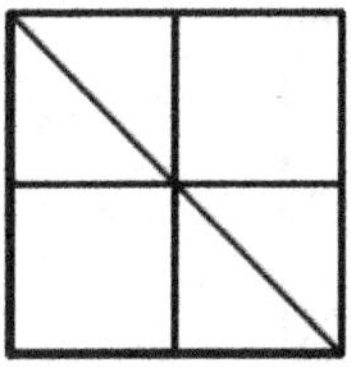

5) What is the sum of all the even factors of 48 that are less than 48?

6) At the class Valentine party 14 students brought candy hearts, 19 brought chocolate kisses, and 7 brought both to give away. If there were 40 students in the class, how many kids did not bring either treat to give away?

*Copyright © 2018 by Perennial Math, Inc. All rights reserved.*

# PERENNIAL MATH

TEST # 4

## February 2018 Rookie Level Answers

| | |
|---|---|
| 1) Two numbers when added together equal 20 and when subtracted equal 6. What is the largest of the two numbers? | 13 |
| 2) Gumballs cost 7 cents and chocolate drops cost 16 cents. If you had to purchase at least one of each, what is the least amount of money you would have to spend if you spent more than 75 cents? | 76 |
| 3) Matthew gave 6 of his 30 marbles to Carl. Now they both have the same number of marbles. They both gave away one fourth of their marbles to Eric. How many marbles did Eric receive? | 12 |
| 4) How many ways can a ladybug crawl from the bottom right corner to the top left corner of this figure if it can only travel up, left or both along the lines? 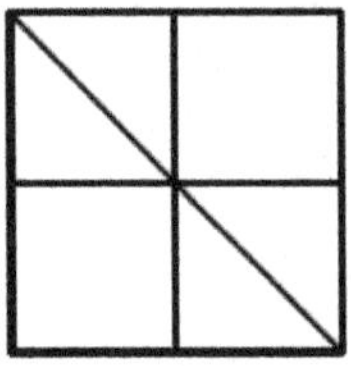 | 11 |
| 5) What is the sum of all the even factors of 48 that are less than 48? | 72 |
| 6) At the class Valentine party 14 students brought candy hearts, 19 brought chocolate kisses, and 7 brought both to give away. If there were 40 students in the class, how many kids did not bring either treat to give away? | 14 |

*Copyright © 2018 by Perennial Math, Inc. All rights reserved.*

PERENNIAL MATH

TEST #4

# February 2018 Rookie Level Solutions

1) Two numbers when added together equal 20 and when subtracted equal 6. To find the largest of the two numbers who sum is 20 and their difference is 6 make a list of numbers that have a sum of 20 and then find the pair whose difference is 6. (20, 0), (19, 1), (18, 2) (17, 3), (16, 4), (15, 5), (14, 6), (13, 7).

The two numbers are 13 and 7. The largest of the two numbers is 13.

**13**

2) Gumballs cost 7 cents and chocolate drops cost 16 cents. If you had to purchase at least one of each. To find the least amount of money you would have to spend if you spent more than 75 cents you could make a table and do a trial and error. The least amount you could spend over 75 cents and buy at least 1 of each candy is 76 cents.

| | | | | | |
|---|---|---|---|---|---|
| Gumballs (.07) | 9 | 8 | 7 | 6 | 4 |
| Chocolate (.16) | 1 | 1 | 2 | 2 | 3 |
| Gumballs total | 0.63 | 0.56 | 0.49 | 0.42 | 0.28 |
| Chocolate total | 0.16 | 0.16 | 0.32 | 0.32 | 0.48 |
| Total money | 0.79 | 0.72 | 0.81 | 0.74 | 0.76 |

**76**

3) If Matthew gave 6 of his 30 marbles to Carl then he would be left with 24 marbles. So they both have the same number of marbles or 24. They both gave away one fourth of their marbles to Eric and ¼ of 24 is 6. So Eric received 6 + 6 = 12 marbles.

**12**

*Copyright © 2018 by Perennial Math, Inc. All rights reserved.*

PERENNIAL MATH

TEST # 4

# February 2018 Rookie Level Solutions

4) How many ways can a ladybug crawl from the bottom right corner to the top left corner of this figure if it can only travel up, left or both along the lines?

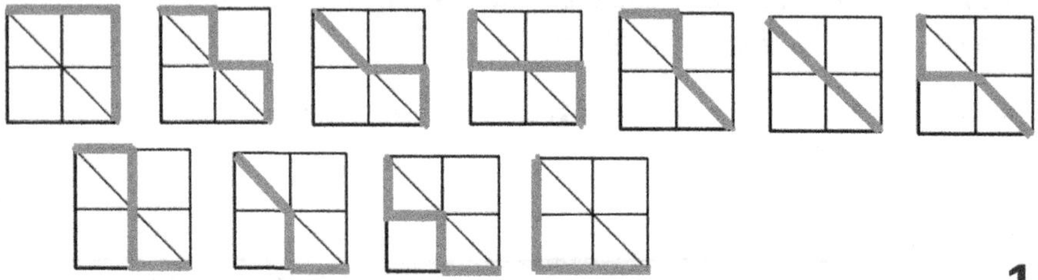

**11**

5) What is the sum of all the even factors of 48 that are less than 48? The factors of 48 are 1, 2, 3, 4, 6, 8, 12, 16, and 24. The sum of the even factors would be 2 + 4 + 6 + 8 + 12 + 16 + 24 = 72

**72**

6) Use a Venn Diagram. At the class Valentine party 14 students brought candy hearts, 19 brought chocolate kisses, and 7 brought both to give away. There were 40 students in the class. To find how many kids did not bring either treat solve for the regions and subtract from the total (40). If 14 students brought candy hearts and 7 brought both, then 14 - 7 = 7 students who only brought the hearts. Likewise, if 19 students brought chocolate kisses and 7 brought both, then 19 – 7 = 12. The three regions are 7 + 7 + 12 = 26. So 40 – 26 = 14 students who did not bring either.

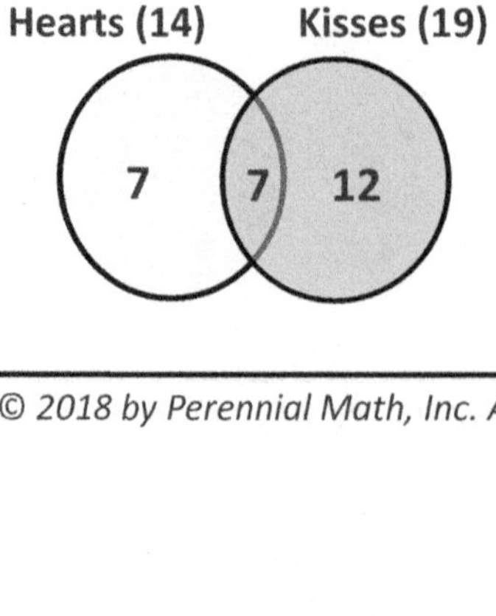


Neither = 40 - (7 + 7 + 12)
= 40 - 26
= 14

**14**

*Copyright © 2018 by Perennial Math, Inc. All rights reserved.*

Name________________________________

# PERENNIAL MATH

TEST #1

## January 2018 Rookie Level

| | |
|---|---|
| 1) 1 + 3 + 5 + 7 + 9 + 11 + 13 + 15 = | |
| 2) John has some cans numbered 1-25. He puts a penny in every can that is divisible by 2, a nickel in every can divisible by 3, a dime in every can divisible by 4, and a quarter in each can divisible by 5. How much total money did he put in the twenty-five cans? | |
| 3) How many triangles of any size are in the picture below? 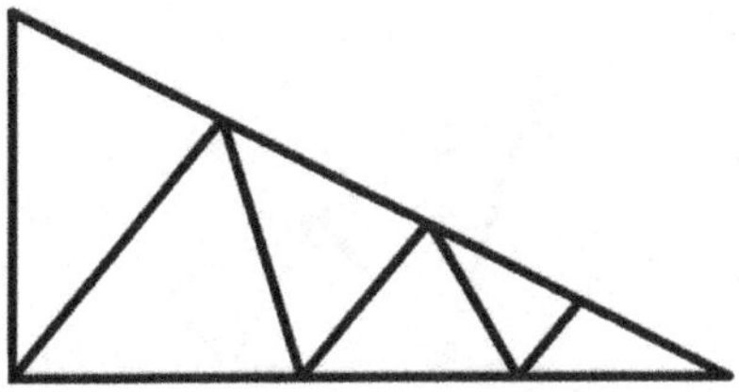 | |
| 4) There were 48 cookies that were shared by the students in Mrs. Brown's class. If each student got 3 cookies and there were 4 more girls than boys, how many boys were in the class? | |
| 5) Mary had some three-leaf clovers and some four-leaf clovers. She counted 30 total plants with 102 total leaves. How many of the plants were four-leaf clovers? | |
| 6) In a survey of 40 students, 17 liked pretzels, 19 liked chips, and 5 liked both. How many students liked neither pretzels nor chips? | |

*Copyright © 2018 by Perennial Math, Inc. All rights reserved.*

# PERENNIAL MATH

TEST # 1

## January 2018 Rookie Level Answers

| | |
|---|---|
| 1) 1 + 3 + 5 + 7 + 9 + 11 + 13 + 15 = | 64 |
| 2) John has some cans numbered 1-25. He puts a penny in every can that is divisible by 2, a nickel in every can divisible by 3, a dime in every can divisible by 4, and a quarter in each can divisible by 5. How much total money did he put in the twenty-five cans? | 2.37 |
| 3) How many triangles of any size are in the picture below? 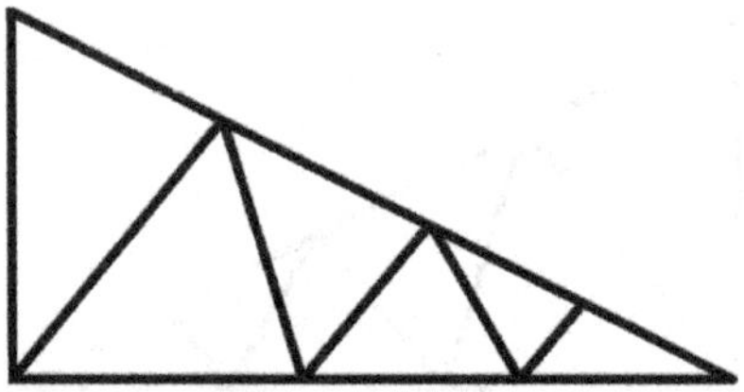 | 11 |
| 4) There were 48 cookies that were shared by the students in Mrs. Brown's class. If each student got 3 cookies and there were 4 more girls than boys, how many boys were in the class? | 6 |
| 5) Mary had some three-leaf clovers and some four-leaf clovers. She counted 30 total plants with 102 total leaves. How many of the plants were four-leaf clovers? | 12 |
| 6) In a survey of 40 students, 17 liked pretzels, 19 liked chips, and 5 liked both. How many students liked neither pretzels nor chips? | 9 |

*Copyright © 2018 by Perennial Math, Inc. All rights reserved.*

PERENNIAL MATH

TEST #1

# January 2018 Rookie Level Solutions

1) Use smart addition by grouping the numbers:

$$1 + 3 + 5 + 7 + 9 + 11 + 13 + 15 =$$
$$(11 + 9) + (13 + 7) + (5 + 15) + (1 + 3) =$$
$$20 + 20 + 20 + 4 = 64$$

**64**

2) List out the numbers that are divisible by 2, 3, 4, and 5 and then count how many of each coin will be put in each can. Add the sums together to find the total amount of money he put in the twenty-five cans.

Can numbers divisible by 2: 2, 4, 6, 8, 10, 12, 14, 16, 18, 20, 22, 24 (12 total cans)
Can numbers divisible by 3: 3, 6, 9, 12, 15, 18, 21, 24 (8 total cans)
Can numbers divisible by 4: 4, 8, 12, 16, 20, 24 (6 total cans)
Can numbers divisible by 5: 5, 10, 15, 20, 25 (5 total cans)
Pennies: 12 cans x .01 = .12
Nickels: 8 cans x .05 = .40
Dimes: 6 cans x .10 = .60
Quarters: 5 cans x .25 = 1.25
He put a total of 31 coins worth .12 + .40 + .60 + 1.25 = $2.37

**2.37**

3) Take a systematic approach to counting the number of triangles.

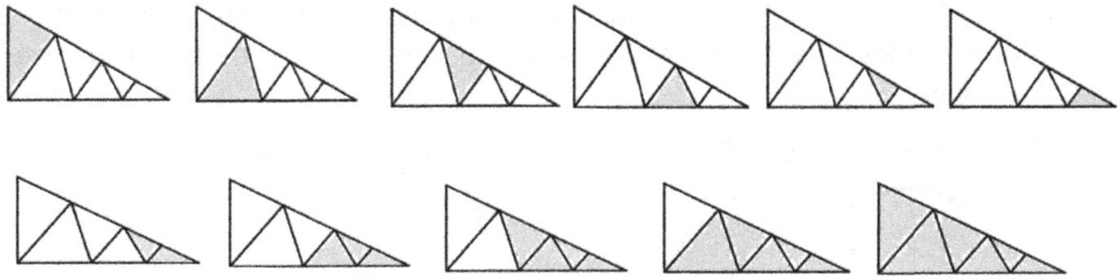

**11**

Copyright © 2018 by Perennial Math, Inc. All rights reserved.

# PERENNIAL MATH

TEST #1

## January 2018 Rookie Level Solutions

4) If there were 48 cookies and each student in Mrs. Brown's class got 3 cookies, then there were a total of 48/3 = 16 students. There were 4 more girls than boys, so 10 + 6 = 16 and 10 is 4 more than 6. Therefore, there were 10 girls and 6 boys in Mrs. Brown's class.

**6**

5) Mary had some three-leaf clovers and some four-leaf clovers. She counted 30 total plants with 102 total leaves. How many of the plants were four-leaf clovers? Make a table and start with same number of each of the plants, 30/2 = 15. Notice that the number of total leaves is greater than 102, so decrease the number of four-leaf clovers until you count 102 total leaves.

| | | | | |
|---|---|---|---|---|
| Three-Leaf Clovers | 15 | 16 | 17 | 18 |
| Four-Leaf Clovers | 15 | 14 | 13 | 12 |
| Total Plants | 30 | 30 | 30 | 30 |
| 3-Leaf Clover Leaves | 45 | 48 | 51 | 54 |
| 4-Leaf Flower Leaves | 60 | 56 | 52 | 48 |
| Total Leaves | 105 | 104 | 103 | 102 |

**12**

6) Use a Venn Diagram. Out of the 40 students, 17 liked pretzels, 19 liked chips and 5 liked both. To find how many people liked neither, solve for the regions and subtract from the total (40). If 17 liked pretzels and 5 liked both, then 17 - 5 = 12 students who only liked pretzels. Likewise, if 19 students liked chips and 5 liked both, then 19 – 5 = 14 who only liked chips. The three regions are 12 + 5 + 14 = 31. So 40 – 31 = 9 students who did not like either.

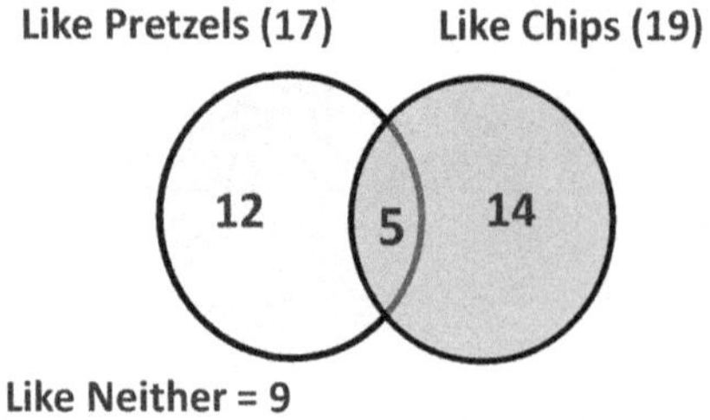


$$40 = 12 + 5 + 14 + n$$
$$40 - 31 = n$$
$$9 = n$$

**9**

*Copyright © 2018 by Perennial Math, Inc. All rights reserved.*

Name______________________________

PERENNIAL MATH

TEST #2

## Febuary 2018 Rookie Level

1) 999 – 111 = 222 + _______

2) The numbers 1 through 4 must appear in each row and column only once. What number will go in the square marked N?

| | | | |
|---|---|---|---|
| | | N | |
| | | | 4 |
| 2 | 3 | | |
| 1 | | 2 | |

3) Bob lined up and he counted and was the $13^{th}$ person from the front of the line and the $14^{th}$ person from the end of the line. How many people were in line?

4) Mandy plants a flower bulb. At the end of the year it produces three extra bulbs. If each bulb produces three extra bulbs at the end of the year, how many bulbs will there be at the end of 5 years?

5) Skip and Brian have 90 marbles together. Skip gave Brian 7 marbles, then they both had the same number of marbles. How many marbles did Brian have at the beginning?

6) How many times does the digit 1 appear between 1 and 100?

*Copyright © 2018 by Perennial Math, Inc. All rights reserved.*

# PERENNIAL MATH

TEST #2

## February 2018 Rookie Level Answers

1) 999 – 111 = 222 + _______

**666**

2) The numbers 1 through 4 must appear in each row and column only once. What number will go in the square marked N?

| | | | |
|---|---|---|---|
| | | N | |
| | | | 4 |
| 2 | 3 | | |
| 1 | | 2 | |

**3**

3) Bob lined up and he counted and was the 13th person from the front of the line and the 14th person from the end of the line. How many people were in line?

**26**

4) Mandy plants a flower bulb. At the end of the year it produces three extra bulbs. If each bulb produces three extra bulbs at the end of the year, how many bulbs will there be at the end of 5 years?

**1,024**

5) Skip and Brian have 90 marbles together. Skip gave Brian 7 marbles, then they both had the same number of marbles. How many marbles did Brian have at the beginning?

**38**

6) How many times does the digit 1 appear between 1 and 100?

**19**

*Copyright © 2018 by Perennial Math, Inc. All rights reserved.*

PERENNIAL MATH

TEST #2

# February 2018 Rookie Level Solutions

1) 999 – 111 = 222 + _______

999 – 111 = 888

So, 888 – 222 = 666

Therefore, 999 – 111 = 222 + 666.

**666**

2) Use logic to solve for N.

Step 1. First you can solve for the missing numbers in the bottom row because numbers 3 and 4 can be placed in the correct squares because there are 3 and 4 in the columns above them.

Step 2. Now move to solve for the two missing numbers 1 and 4 in the third row. The number 4 in row 2 above will help solve for their placement.

Step 3. Next you may fill in the remaining number 2 in the top of column 4. Then move on to filling in the missing numbers 3 and 4 in column 1. The number 4 in column 4 will help you decide that placement.

Step 4. Finally, you will be able to use the numbers in the surrounding columns and rows to complete the puzzle. The missing number N is 3.

Step 1

| | | | |
|---|---|---|---|
| | | | |
| | | | 4 |
| 2 | 3 | | |
| 1 | 4 | 2 | 3 |

Step 2

| | | | |
|---|---|---|---|
| | | | |
| | | | 4 |
| 2 | 3 | 4 | 1 |
| 1 | 4 | 2 | 3 |

Step 3

| | | | |
|---|---|---|---|
| 4 | | | 2 |
| 3 | | | 4 |
| 2 | 3 | 4 | 1 |
| 1 | 4 | 2 | 3 |

Step 4

| | | | |
|---|---|---|---|
| 4 | 1 | 3 | 2 |
| 3 | 2 | 1 | 4 |
| 2 | 3 | 4 | 1 |
| 1 | 4 | 2 | 3 |

**3**

3) Bob lined up and he counted and was the $13^{th}$ person from the front of the line and the $14^{th}$ person from the end of the line. Draw a picture to see how many people were in line. There were 26 people in the line.

**26**

Copyright © 2018 by Perennial Math, Inc. All rights reserved.

PERENNIAL MATH

TEST # 2

# February 2018 Rookie Level Solutions

**4) Mandy plants a flower bulb. At the end of the year it produces three extra bulbs. If each bulb produces three extra bulbs at the end of the year, how many bulbs will there be at the end of 5 years? Make a chart. There were 1,024 bulbs at the end of 5 years.**

| | Year 1 | Year 2 | Year 3 | Year 4 | Year 5 |
|---|---|---|---|---|---|
| Bulbs beginning | 1 | 4 | 16 | 64 | 256 |
| Bulbs produced | 3 | 12 | 48 | 192 | 768 |
| Total bulbs | 4 | 16 | 64 | 256 | 1024 |

**1,024**

**5) Skip and Brian have 90 marbles together. Skip gave Brian 7 marbles, then they both had the same number of marbles. How many marbles did Brian have at the beginning? Work backward and divide 90 by 2 see how many marbles they both ended up with. 90/2 = 45. If they ended up with 45 each and Brian received 7 marbles from Skip, then 45 – 7 = 38. Brian had 38 marbles at the beginning.**

**38**

**6) How many times does the digit 1 appear between 1 and 100? Make a list of the numbers *BETWEEN* 1 and 100 that have the digit 1 in them:**
**10, 11, 12, 13, 14, 15, 16, 17, 18, 19 = 11 times (be sure to count both ones in 11)**
**21, 31, 41, 51, 61, 71, 81, 91 = 8 times**

**The digit 1 appears 19 times between 1 and 100.**

**19**

*Copyright © 2018 by Perennial Math, Inc. All rights reserved.*

Name________________________________

PERENNIAL MATH

TEST #3

## March 2018 Rookie Level

1) Complete the sequence: ( 1, 5, 2, 6, 3, ___, ____, ____). What is the sum of the three missing numbers?

2) What is the largest two-digit number that has the same answer when you add or subtract the two digits?

3) John has 23 coins which are only dimes or quarters. He counted his money and found out he had $4.55. How many dimes did he have?

4) Howie woke up at 4:56 and noticed that the numbers on his digital clock were ascending consecutive numbers. How many total minutes will pass before the clock shows all consecutive numbers going up again? Express your answer in total minutes.

5) 1111 is a 4-digit number whose digits add up to 4 (1 + 1 + 1 + 1 = 4). How many total 4-digit numbers have a digit sum of 4?

6) Mike cut a log into 3 pieces in 12 minutes. How many minutes will it take him to cut a log into 4 pieces if he is cutting at the same rate?

*Copyright © 2018 by Perennial Math, Inc. All rights reserved.*

PERENNIAL MATH

TEST #3

# March 2018 Rookie Level Answers

| | |
|---|---|
| 1) Complete the sequence: ( 1, 5, 2, 6, 3, ___, ____, ____). What is the sum of the three missing numbers? | 19 |
| 2) What is the largest two-digit number that has the same answer when you add or subtract the two digits? | 90 |
| 3) John has 23 coins which are only dimes or quarters. He counted his money and found out he had $4.55. How many dimes did he have? | 8 |
| 4) Howie woke up at 4:56 and noticed that the numbers on his digital clock were ascending consecutive numbers. How many total minutes will pass before the clock shows all consecutive numbers going up again? Express your answer in total minutes.   | 458 |
| 5) 1111 is a 4-digit number whose digits add up to 4 (1 + 1 + 1 + 1 = 4). How many total 4-digit numbers have a digit sum of 4? | 20 |
| 6) Mike cut a log into 3 pieces in 12 minutes. How many minutes will it take him to cut a log into 4 pieces if he is cutting at the same rate? | 18 |

*Copyright © 2018 by Perennial Math, Inc. All rights reserved.*

# PERENNIAL MATH

TEST # 3

## March 2018 Rookie Level Solutions

1) Complete the sequence: ( 1, 5, 2, 6, 3, ___, ___, ___). What is the sum of the three missing numbers?
The pattern is + 4, - 3, so the next three numbers would be
(3 + 4) = 7; (7 – 3) = 4; (4 + 4) = 8.
The sum of the three missing numbers is
7 + 4 + 8 = 19.

**19**

2) What is the largest two-digit number that has the same answer when you add or subtract the two digits? The only two digit numbers that have the same answer when you add or subtract the two digits are numbers that end with 0: 10, 20, 30, 40, 50, 60, 70, 80, 90. The largest two-digit number is 90.

**90**

3) John has 23 coins which are only dimes or quarters which total $4.55. To find out how many dimes he had make a table. Start with half of each coin. With 11 quarters and 12 dimes, you can see that he does not have enough money and needs to increase the amount of quarters that he has so add quarters and take away dimes until you reach the target amount of $4.55. He had 15 quarters and 8 dimes.

| Quarters | Dimes | Total Coins | Amount |
|---|---|---|---|
| 11 | 12 | 23 | 3.85 |
| 12 | 11 | 23 | 4.10 |
| 13 | 10 | 23 | 4.25 |
| 14 | 9 | 23 | 4.40 |
| 15 | 8 | 23 | 4.55 |

**8**

*Copyright © 2018 by Perennial Math, Inc. All rights reserved.*

PERENNIAL MATH

TEST #3

# March 2018 Rookie Level Solutions

4) There are no numbers in single digits that follow after 4 that will allow for ascending numbers because 56 is the highest ascending amount that the minutes can be. The next number that can appear is 12:34 because all numbers are going up. So the difference in time is 12:34 – 4:56 which is 7 hours and 38 minutes. That time converted to all minutes is (7 x 60) + 38 = 458 minutes.

**458**

5) 1111 is a 4-digit number whose digits add up to 4 (1 + 1 + 1 + 1 = 4). To find out how many total 4-digit numbers have a digit sum of 4 make an organized list. There are 20 different 4-digit numbers with a digit sum of 4.

| | | | |
|---|---|---|---|
| 1003 | 2002 | 3001 | 4000 |
| 1012 | 2011 | 3010 | |
| 1021 | 2020 | 3100 | |
| 1030 | 2101 | | |
| 1102 | 2110 | | |
| 1111 | 2200 | | |
| 1120 | | | |
| 1201 | | | |
| 1210 | | | |
| 1300 | | | |

**20**

6) Act this out. If Mike cut a log into 3 pieces in 12 minutes, he had to make only two cuts because 1 cut yields 2 pieces and 2 cuts would yield 3 pieces. So it took 12 minutes/2 cuts = 6 minutes for each cut. Cutting at the same rate, it will take him 3 cuts x 6 minutes or 18 minutes to cut 4 pieces.

1 cut equal 2 pieces

**18**

*Copyright © 2018 by Perennial Math, Inc. All rights reserved.*

Name________________________________

PERENNIAL MATH

TEST #4

# April 2018 Rookie Level

1) How many ways can three girls line up for lunch?

2) In the sequence 5, ..., ___, 46, 75 two consecutive entries are added to get the next entry. The last two entries are 46 and 75. The first entry is 5. How many entries are there in the entire sequence?

3) Use the information from each scale to solve for the value that belongs in the last scale. How much do three stars weigh?

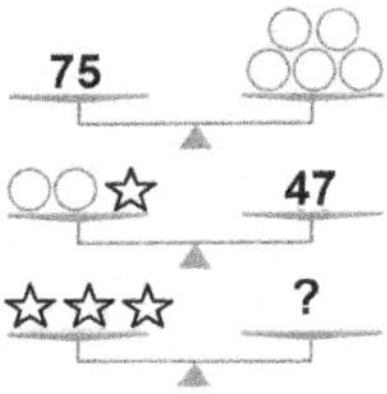


4) Queen Elizabeth was born in April 1926 and became queen February 1952. How old will she be when they celebrate her 70th year as queen?

5) Mike and Bill are cousins whose ages are the same two different digits, only reversed. The sum of their ages is 132. Mike is less than 20 years older than Bill. How old is Bill?

6) Holiday stickers costs $2.75 per package of 250 stickers. If Carolyn buys 4500 stickers with her $50, how much money will she get back? Write your answer in decimal form.

*Copyright © 2018 by Perennial Math, Inc. All rights reserved.*

PERENNIAL MATH

TEST #4

# April 2018 Rookie Level Answers

| | |
|---|---|
| 1) How many ways can three girls line up for lunch? | 6 |
| 2) In the sequence 5, …, ___, 46, 75 two consecutive entries are added to get the next entry. The last two entries are 46 and 75. The first entry is 5. How many entries are there in the entire sequence? | 6 |
| 3) Use the information from each scale to solve for the value that belongs in the last scale. How much do three stars weigh? 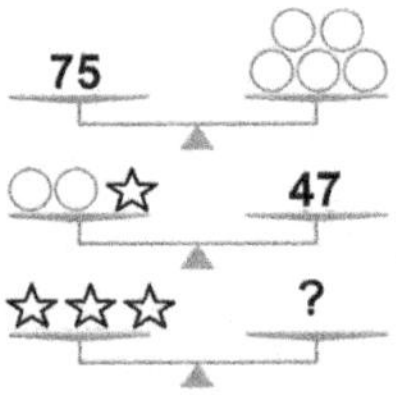  | 51 |
| 4) Queen Elizabeth was born in April 1926 and became queen February 1952. How old will she be when they celebrate her 70th year as queen? | 95 |
| 5) Mike and Bill are cousins whose ages are the same two different digits, only reversed. The sum of their ages is 132. Mike is less than 20 years older than Bill. How old is Bill? | 57 |
| 6) Holiday stickers costs $2.75 per package of 250 stickers. If Carolyn buys 4500 stickers with her $50, how much money will she get back? Write your answer in decimal form. | .50 |

*Copyright © 2018 by Perennial Math, Inc. All rights reserved.*

# PERENNIAL MATH

TEST #4

## April 2018 Rookie Level Solutions

1) How many ways can three girls line up for lunch? Make an organized list of possibilities for girl A, B, and C. ABC, ACB, BAC, BCA, CAB, CBA. There are 6 possibilities for the arrangement of the girls.

**6**

2) In the sequence 5, …, __, 46, 75 two consecutive entries are added to get the next entry. The last two entries are 46 and 75. The first entry is 5. Work backward to find out how many entries are there in the entire sequence. 75 – 46 = 29. Next 46 – 29 = 17. Then 29 – 17 = 12. Finally 17 – 12 = 5. So the sequence is 5, 12, 17, 29, 46, 75. There are 6 entries in the sequence.

**6**

3) How much do three stars weigh? The first step is to solve for the circles. 5 circles weigh 75 so 75 / 5 = 15. Two circles plus one star weigh 47. We know that 2 circles weigh 30 so a star must weigh 47 – 30 = 17. Therefore, 3 stars weigh 3 x 17 or 51.

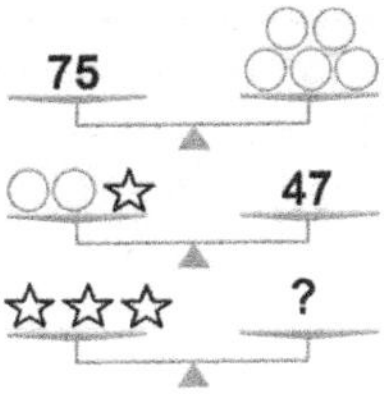


○ = 15

2(15) + ☆ = 47, so the star = 17

3(17) = 51

**51**

*Copyright © 2018 by Perennial Math, Inc. All rights reserved.*

PERENNIAL MATH

TEST #4

## April 2018 Rookie Level Solutions

4) Queen Elizabeth was born in April 1926 and became queen February 1952. If she became queen in 1952, then 1952 + 70 years = 2022. If she was born in 1926 then she is 2022 – 1926 = 96 years old. Since she became queen before her actual birthday, that would make her 95 years old instead of 96. She will be 95 when they celebrate her 70$^{th}$ year as queen.

**95**

5) Mike and Bill's ages are the same two digits, only reversed and the sum of their ages is 132. Mike is less than 20 years older than Bill. To find out how old Bill is we first have to get a starting point. If we subtract 132 – 20 = 112 we can get what the age is going to be around with the twenty year difference. So divide this in half to get a starting point. 112/2 = 56. Now 56 + 76 = 132, but Mike is less than 20 years older than Bill and Bill has to be older. So, go up by one year and 57 + 75 = 132. This works because the digits are reversed and the sum is 132. Therefore, Bill is 57 years old.

**57**

6) Holiday stickers costs $2.75 per package of 250 stickers. To find out how much money Carolyn will have to spend if she wants to buy 4500 stickers, first find out how many packs of stickers she must buy. Divide 4500 by 250. 4500/250 = 18. So If she buys 18 packages at $2.75 per package, she will spend $49.50. $50.00 - $49.50 = .50

**.50**

*Copyright © 2018 by Perennial Math, Inc. All rights reserved.*